The Complete Cannabis Cloning Guide

Chapter 1: Introduction

In the captivating world of cannabis cultivation, the concept of cannabis cloning stands as a horticultural marvel. At its essence, cloning is the art of replicating a genetically identical copy of a cannabis plant, an intricate process that has captivated the imaginations of cultivators for generations. This book begins by unfurling the enigmatic allure of cannabis cloning, inviting readers into a realm where botanical science and cultivation expertise intertwine.

The Importance of Cloning in Cannabis Cultivation

Cannabis cloning is not merely a facet of cultivation; it's a keystone technique, revered and relied upon by growers of all calibers. It serves as the cornerstone for consistency in cannabis production. By propagating genetically identical plants, cultivators can perpetuate the prized traits of their favorite strains, be it remarkable potency, exquisite flavors, or prolific yields. Cloning ensures that the progeny retain the exact genetic blueprint of their parent, offering a level of predictability that is the holy grail of cannabis cultivation.

But the significance of cloning extends far beyond the realm of genetics. It's also a strategy for resource efficiency. Growers can save both time and resources by cultivating from clones rather than seeds, bypassing the variability that seeds inherently introduce.

Furthermore, cannabis cloning allows for the preservation of rare or endangered strains that might otherwise be lost to the annals of botanical history. It's a conservation tool that safeguards genetic diversity in the cannabis world.

A Glimpse of What to Expect in the Book

As you journey deeper into the pages of this book, you'll find a treasure trove of knowledge and wisdom waiting to be unearthed. It's a comprehensive guide that navigates the complexities of cannabis cloning with precision and clarity. Each chapter unfolds a new layer of understanding, from selecting the perfect mother plant to nurturing clones into thriving cannabis plants. Within these chapters, you'll discover meticulous techniques, expert tips, and a wealth of insights to elevate your cloning endeavors. Whether you're a novice cultivator or an experienced horticulturalist, this book promises to equip you with the skills and expertise needed to master the art of cannabis cloning. It's a roadmap to unlocking the full potential of your cannabis garden, ensuring consistency, potency, and quality in every harvest.

In closing, this book is an invitation to embark on a journey of cultivation, discovery, and horticultural mastery. It beckons you to delve deeper into the intricacies of cannabis cloning, to explore the nuances of selection, propagation, and care, and to emerge as a cultivator who can craft not just a garden but an artful masterpiece..

Chapter 2: Selecting the mother Plant

In the vast and intricate world of cannabis cloning, Chapter 2 stands as a cornerstone, illuminating the path to the heart of successful propagation - selecting the perfect mother plant. With a keen eye for detail, this chapter provides an indispensable roadmap, offering a level of guidance that is as precise as it is comprehensive.

At its core, this chapter delves into the art and science of choosing the right mother plant. It elevates the selection process beyond mere chance or whim, offering readers a systematic approach that ensures the genetic foundation of their cannabis cultivation is rock-solid. It's a process that involves careful scrutiny, evaluating not just the immediate attributes of the plant, but also its potential to yield a lineage of vibrant, desirable offspring.

The quest for the perfect mother plant in cannabis cloning is akin to the pursuit of a botanical masterpiece, a harmonious blend of science and artistry. The perfect mother plant embodies a delicate balance of desirable traits, health, and vigor. In the intricate world of cannabis cultivation, this elusive ideal takes on multifaceted dimensions.

First and foremost, the perfect mother plant is a treasure trove of desirable traits. These traits might include high potency, distinguished terpene profiles, specific flavors, or the ability to produce robust yields. It's a living embodiment of the cultivator's aspirations, a botanical canvas onto which one can paint their vision of the ideal cannabis strain. In the world of cloning, these traits are not mere luxuries but the essential building blocks of success. Equally important is the mother plant's overall health and condition. This foundational aspect ensures that the genetic heritage passed down to the clone offspring is one of vitality and vigor. A healthy mother plant is free from the insidious grip of pests and diseases. It stands tall, an exemplar of robust physiology, ready to transmit its vitality to the next generation. It's the cornerstone upon which successful cloning endeavors are built.

Furthermore, the perfect mother plant boasts resilience and stability. It thrives in the chosen growing environment, adapting to fluctuations in light, temperature, and humidity with grace. This resilience is essential, ensuring that the mother plant can withstand the demands of repeated cloning cycles without degradation in the quality of clone offspring.

In the art of cannabis cloning, the perfect mother plant is also a story of age and experience. It's not a fledgling plant but one that has matured and proven its worth over time. These veteran plants often produce more consistent and reliable clones, solidifying their position as prized assets in the cultivator's arsenal.

In conclusion, the perfect mother plant in the realm of cannabis cloning is the embodiment of aspiration, health, resilience, and experience. It's a botanical masterpiece waiting to be discovered, cultivated, and revered. Cultivators who master the art of selecting and nurturing such mother plants are poised to unlock the full potential of cannabis cloning, crafting strains that reflect their unique vision and expertise.

Chapter 3: Preparing Your Cloning Tools

Chapter 3 of our cannabis cloning guide unveils the critical tools that form the backbone of successful cannabis propagation. These tools, often overlooked, are the practical instruments that turn theory into reality, and in this chapter, we'll explore their individual roles and their collective significance.

Firstly, the razor or scalpel serves as a cultivator's precision tool, enabling clean and accurate cuts on chosen cannabis plants, minimizing stress on the cuttings. Cloning gel, akin to the elixir of life for your cuttings, plays a vital role by jumpstarting root development, significantly enhancing the chances of successful cloning. Rooting cubes or suitable cloning mediums provide a nurturing environment during the early growth stages, offering stability, moisture retention, and the ideal conditions for root formation.

Humidity domes create a microclimate of consistent moisture, a critical factor in clone success. They prevent excessive moisture loss, ensuring an optimal environment for root development. Lastly, spray bottles might seem unassuming but are indispensable for maintaining humidity levels. Regular misting ensures a consistent humid atmosphere, safeguarding your cuttings from drying out.

The overarching importance of maintaining sterility looms large throughout this chapter. Sterility is the guardian of your cloning success, preventing any contamination that could compromise your clones. It safeguards the rooting hormone and cuttings from harmful pathogens. Moreover, it prolongs the lifespan of your tools, making cleanliness a cost-saving ally.

In summary, Chapter 3 underscores the vital role of these tools and the paramount importance of sterility. As you embark on your cloning journey, remember that precision and cleanliness are your trusted companions, ensuring the health and vitality of your cannabis clones.

Chapter 4: Taking the Cutting

Taking a cutting from a cannabis plant is a pivotal step in the cloning process, and its precise execution is critical for successful propagation. To master this fundamental skill, follow these detailed steps. Start by judiciously choosing a branch or shoot from the mother plant that exhibits the desired traits you wish to replicate in your clone. Opt for branches that are robust, healthy, and devoid of pests or diseases, as these attributes will be inherited by your clone.

Next, pinpoint a node on the chosen branch. Nodes are crucial to cloning, housing specialized cells with the remarkable ability to transform into roots under favorable conditions. This is where root growth will originate in your cutting. Ensuring sterility, prepare your tools, including a sharp, sterilized razor or scalpel. Sterilization is essential to prevent contamination, safeguarding the health of your cutting.

With your tools at the ready, make a precise cut just below the identified node. A clean, diagonal cut is crucial, and the length of your cutting matters too. Aim for a length of approximately 4 to 6 inches. This length provides an optimal balance, offering enough stem for stability while avoiding excessive length that might hinder the cutting's ability to support its growth.

Explaining the Significance of Nodes

Nodes are the powerhouses of plant growth. They house specialized cells called meristematic cells, which possess the extraordinary capability to differentiate into roots, branches, or leaves. When you make a cutting just below a node, you're tapping into this inherent potential for root development. Nodes serve as the launchpad for new growth, enabling your cutting to establish a robust root system and eventually flourish into a healthy, self-sufficient plant.

Ensuring the Proper Length and Nodes on the Cutting

The length of your cutting is pivotal to its success. A cutting that is excessively short may lack the necessary energy reserves to establish itself, while one that is excessively long could struggle to support its own growth. Strive for a length ranging from 4 to 6 inches, striking a harmonious balance.
Equally essential is the inclusion of at least one node on your cutting. Nodes are the epicenters of growth, where roots will sprout and fresh branches will emerge. Without nodes, your cutting lacks the foundational building blocks

needed for its development. In essence, the significance of nodes cannot be overstated in the art of cannabis cloning, as they hold the key to a thriving, flourishing plant.

Chapter 5: Removing Lower Leaves

Leaf removal, a critical step in the cloning process of cannabis, is not just a matter of aesthetics but a key practice with profound implications for the health and vitality of your clones. In this section, we will delve into why leaf removal is paramount and how it directly contributes to the success of your cannabis cloning endeavors. Moisture Retention: One of the primary reasons for removing lower leaves during cloning is to enhance moisture retention. Leaves are like tiny factories, continually transpiring water vapor into the atmosphere through tiny openings called stomata. When you remove the lower leaves, you reduce the surface area from which moisture can escape. This effectively creates a microenvironment with higher humidity around the cutting. Maintaining adequate humidity is crucial, as it prevents the cutting from drying out before it can develop its own roots. By minimizing moisture loss through transpiration, you provide your clone with a better chance of survival during this critical phase.

Energy Allocation: Lower leaves also play a role in the allocation of energy within the cutting. These leaves require energy for maintenance, including water uptake and nutrient transport. When you remove them, the cutting can redirect its limited energy resources towards root development. In essence, it shifts the plant's focus from maintaining unnecessary foliage to fostering the growth of a strong, healthy root system. This reallocation of energy enhances the cutting's chances of successfully establishing itself as a new, independent plant.

Pruning Techniques When Discussing Cloning Cannabis: There are several techniques you can employ when removing lower leaves during the cloning process. These include:

- Simple Leaf Removal: The most common method is to gently remove the lower leaves by pinching or cutting them off using clean, sterilized scissors or pruning shears. Ensure that you leave a few leaves at the top of the cutting to aid in photosynthesis and provide some energy for the clone during its initial stages.

- Topping: Topping involves the removal of the top part of the cutting, which not only reduces the overall size of the cutting but also encourages bushier growth as multiple shoots emerge from the point of removal.

- FIM (Fuck I Missed)**: Similar to topping, FIMming involves removing the tip of the cutting. However, it is done less aggressively, often by pinching or cutting off around 80% of the tip. This technique can also promote bushier growth and can be particularly useful when you want to encourage lateral branching.

In conclusion, leaf removal is a pivotal practice in the art of cannabis cloning. It aids in moisture retention and energy allocation, creating the ideal conditions for the cutting to develop roots and establish itself as a healthy, thriving clone. Understanding the importance of this step and employing effective pruning techniques are essential for successful cannabis cloning endeavors.

Chapter 6: Applying Rooting Hormone

Rooting hormone is a crucial ally in the art of cannabis cloning, holding the key to transforming a mere cutting into a flourishing, independent plant. Its fundamental function lies in stimulating the growth of roots in plant cuttings, expediting the process of establishing a new, self-sustaining plant. This section unveils the profound significance of rooting hormone, presents different types of rooting hormones, and highlights how to apply them effectively.

Rooting Hormone Significance:

Rooting hormone serves as a potent growth regulator, specifically designed to kickstart root development in plant cuttings. Its primary role is to encourage the emergence of roots at the base of the cutting, facilitating a swift and efficient transition to a thriving, self-sustaining plant. Without rooting hormone, root development may still occur, but it's often slower and less predictable. The application of rooting hormone significantly enhances the likelihood of successful cloning by jumpstarting this crucial phase.

Different Types of Rooting Hormones:

Various types of rooting hormones are available, each with its unique formulation. The principal types include powder, gel, liquid, and even natural alternatives like aloe vera.

Powder rooting hormone is a common choice, offering simplicity in application. Gel, on the other hand, provides a thicker consistency, reducing mess during use.

Liquid rooting hormone, when properly diluted, serves as a soaking solution for cuttings, although it requires careful handling to ensure the cutting's base is adequately saturated.

Additionally, aloe vera has emerged as a reliable natural rooting hormone, harnessing its inherent growth-promoting properties to facilitate root development in cuttings.

How to Apply Rooting Hormone Effectively:

To harness the full potential of rooting hormone, follow these steps for effective application:

- Prepare the Cutting: Begin by trimming the cutting to the desired length, ensuring it includes at least one node—the epicenter for root formation.
- Dip or Coat: For powder or gel rooting hormone, immerse the cut end of the clone into the rooting hormone, ensuring thorough coverage. When using liquid hormone, briefly submerge the cutting's base in the solution.
- Shake Off Excess: To avoid excess hormone, gently tap or shake the cutting, leaving a thin, even coating.
- Plant in Cloning Medium: Insert the treated cutting into your chosen cloning medium, such as rooting cubes or aloe vera gel,

creating a suitable hole to accommodate the clone.

In summary, rooting hormone acts as a catalyst in the cloning process, propelling root development and increasing the likelihood of successful propagation. Various types, including powder, gel, liquid, and natural alternatives like aloe vera, are available to cater to different preferences. When applying rooting hormone, ensure comprehensive coverage to optimize its effectiveness, unlocking the full potential of your cannabis clones.

Chapter 7: Planting the Cutting

Selecting the appropriate cloning medium is a pivotal decision when embarking on cannabis cloning. The cloning medium serves as the nurturing cradle for your cannabis cutting during its early growth stages, playing a foundational role in the health and vitality of your clones. This section underscores the importance of making the right choice, outlines the step-by-step process of planting the cutting, and emphasizes the critical need to ensure nodes are properly covered.

The Significance of Cloning Medium:

The cloning medium is more than just a vessel; it's a vital component that directly impacts the success of your cloning efforts. It provides essential stability, moisture retention, and support for the initial stages of root development. The type of cloning medium you choose can have a significant influence on the outcome of your cloning endeavor. Options such as rockwool, peat pellets, coco coir, and aloe vera gel each offer unique advantages. Your selection may depend on factors like personal preference, availability, and the specific requirements of the cannabis strains you're working with.

The Process of Planting the Cutting:

- Prepare the Medium: Before planting, ensure your chosen cloning medium is adequately prepared. It should be moist but not excessively saturated. Follow the manufacturer's instructions for proper hydration.
- Create a Hole: Using a clean and sterilized tool, like a dibber or pencil, gently create a small hole or depression in the cloning medium. This hole should be deep enough to accommodate the cutting without damaging the delicate root tip.
- Plant the Cutting: Delicately insert the treated cutting into the prepared hole within the cloning medium. It's paramount to ensure that the node, which is the juncture where leaves meet the stem, is

completely submerged in the medium. The node is the epicenter for root development.

- Secure the Cutting: Carefully but firmly pack the cloning medium around the cutting to provide stability and to ensure optimal contact between the medium and the stem. Handle the cutting gently to avoid any harm.

- Maintain Proper Spacing: If you're planting multiple cuttings within the same medium, be mindful of spacing. Adequate spacing prevents overcrowding, which can impede airflow and increase the risk of mold or disease transmission.

Ensuring Nodes Are Properly Covered:

A key aspect of planting the cutting is guaranteeing that the nodes are fully covered by the cloning medium. Nodes serve as the vital hubs for root development, and by ensuring they're submerged, you promote the emergence of robust, healthy roots. An exposed node might struggle to develop roots, potentially compromising the overall success of your cloning efforts.

In conclusion, the choice of a suitable cloning medium is foundational to the success of your cannabis cloning endeavor. The process of planting the cutting involves creating a proper hole, gently inserting the cutting, and ensuring that the node is entirely covered by the medium. These meticulous steps collectively create an environment conducive to the thriving growth of your clones, fostering their successful transition from mere cuttings to robust, independent cannabis plants.

Chapter 8: Creating a Favorable Environment

Successful cannabis cloning hinges on creating an optimal environment for your cuttings to thrive. Three crucial factors in this endeavor are maintaining high humidity levels, providing the right temperature and lighting, and understanding the significance of misting for humidity control.

Maintaining High Humidity Levels:

High humidity is a non-negotiable prerequisite for the success of cannabis cuttings during the cloning process. This is because cuttings lack an established root system to absorb moisture. High humidity, typically around 70-75%, creates a protective shield of moisture around the cutting. This shield ensures that the cutting doesn't lose more moisture through transpiration than it can absorb through its leaves. This microclimate of moisture enveloping the cutting encourages the development of roots without the risk of dehydration.

Appropriate Temperature and Lighting:

Temperature and lighting are equally pivotal in cannabis cloning. Maintaining a temperature range of 70-75°F (21-24°C) creates an ideal climate for your cuttings to focus on root development. Below this range, growth slows, while above it, stress and disease susceptibility increase. Lighting, although different from the needs of mature plants, is also crucial. While many growers prefer to keep cuttings in a shaded area, it's beneficial to provide gentle, diffused light. This encourages the cutting to perform photosynthesis, albeit at a reduced rate compared to mature plants. Grow lights, such as fluorescents or LEDs, are often used to ensure consistent, low-intensity lighting, minimizing stress on the cuttings.

The Importance of Misting for Humidity:

Misting is a fundamental practice in maintaining high humidity levels around your cannabis cuttings. It involves delicately spraying the surrounding air and the cuttings themselves with water. Misting serves several key purposes:

- Preventing Dehydration: Misting helps counteract moisture loss through transpiration, a vital function for cuttings without a developed root system. Regular misting ensures they remain well-hydrated.
- Creating a Humid Microclimate: The fine mist generated in the growing environment establishes a microclimate of high humidity

around the cuttings. This microclimate acts as a protective shield, safeguarding against desiccation.

- Enhancing Gas Exchange: Misting also supports essential gas exchange. By moistening the air, you facilitate the absorption of carbon dioxide, which is crucial for photosynthesis, and the release of oxygen.

To implement misting effectively, employ a clean spray bottle filled with water, and lightly mist the cuttings and the surrounding air on a regular basis. Caution is advised not to overdo it, as excessive moisture can lead to mold and other issues.

In summary, creating the right conditions is fundamental for the successful cloning of cannabis. High humidity levels, suitable temperature, and gentle lighting are essential for your cuttings to concentrate on root development and become healthy, independent plants. Misting plays a pivotal role in this process, ensuring that your cuttings remain adequately hydrated and shielded from dehydration during their early growth stages

Chapter 9: Monitoring and Caring for the Clones

Ensuring the Health of Your Clones:

Regular Inspections, Humidity, Light, and Water
Maintaining the health and vitality of your cannabis clones demands meticulous attention to detail and a commitment to providing the ideal conditions for their growth. This section delves into the essential aspects of ensuring your clones thrive: conducting regular inspections for wilting, pests, and diseases; preserving optimal

humidity levels; and furnishing adequate light and water.
Regular Inspections for Wilting, Pests, and Diseases:
Vigilance is paramount when it comes to the well-being of your clones. Regular inspections, ideally performed daily, serve as a critical preventive measure against wilting, pests, and diseases. Wilting can be an early sign of stress or inadequate hydration, and swift action is crucial. Similarly, pests, such as aphids or spider mites, can quickly infest your clones, wreaking havoc on their health. Disease, too, can strike swiftly, especially in high humidity environments. By conducting routine inspections, you can detect and address these issues promptly, increasing the chances of successful cloning.

Maintaining Humidity Levels:

Humidity control is a constant concern during the cloning process. High humidity, typically maintained around

70-75%, is vital for cuttings lacking a developed root system. It ensures they don't lose more moisture through transpiration than they can absorb through their leaves. Maintaining these levels involves meticulous monitoring and adjustments as needed. Utilizing a humidity dome or propagator can help create a stable microenvironment for your clones, but regular checks with a hygrometer are necessary to ensure humidity remains within the optimal range.

Providing Adequate Light and Water:

Light and water are two life-sustaining elements for your clones. Although they require reduced light compared to mature plants, providing some gentle, diffused light is essential. Fluorescent or LED grow lights are commonly employed to offer consistent, low-intensity lighting that minimizes stress on the cuttings. Adequate watering is equally vital. While you must maintain high humidity, avoid waterlogged conditions that can promote disease. A careful balance is required, ensuring the medium remains moist but not soggy. Remember, the absence of a developed root system means cuttings rely heavily on moisture from the air and their immediate surroundings.

In conclusion, maintaining the health and prosperity of your cannabis clones involves a multi-faceted approach. Regular inspections to detect wilting, pests, and diseases are indispensable. Humidity control within the optimal range, typically around 70-75%, is fundamental for cuttings' success. Providing adequate light, albeit at reduced intensity, and water is equally pivotal. By diligently addressing these aspects, you'll foster an environment where your clones can flourish, establishing themselves as robust, self-sufficient cannabis plants.

Chapter 10: Transplanting and Providing Ongoing Care

Transition to Larger Containers or Growing Mediums

As your cannabis clones begin to develop roots and grow into small plants, a pivotal moment arises when they outgrow their initial cloning environment. This transition to larger containers or growing mediums is a crucial step in the cloning process. It marks the beginning of their journey towards becoming mature, productive cannabis plants.

When your clones have established a healthy root system, usually within 1-3 weeks, it's time to move them to larger containers or a suitable growing medium. This step allows the young plants more space for root expansion and access to a richer source of nutrients. The choice of container or medium largely depends on your cultivation method and preferences. Common options include transplanting into larger pots filled with quality potting soil, hydroponic systems, or outdoor planting in well-prepared soil.

Continued Care for Growing Clones

As you make the transition, it's essential to maintain attentive care for your growing clones. Continue to monitor them for signs of stress, wilting, pests, or diseases. Keep humidity levels within the appropriate range, adjusting as necessary for their new environment. Proper watering is key during this stage. Be cautious not to overwater or let the medium become waterlogged, as young plants are more susceptible to root rot in overly wet conditions.
Additionally, ensure they receive suitable lighting. If you're transitioning

from an indoor cloning setup, gradually acclimate the plants to the intensity of your grow lights to prevent shock. Adequate light exposure is crucial for healthy photosynthesis and overall growth.

Preparing for the Next Stages of Growth

The transition to larger containers or mediums marks a significant milestone, but it's not the end of the journey. It's a preparation for the next stages of growth. At this point, your clones are still in their vegetative phase, focusing on building a robust root system and green, leafy growth. As they continue to flourish, you'll eventually need to make decisions about when and how to induce flowering for the production of buds.

To prepare for this next phase, consider factors like the desired plant size, strain characteristics, and available growing space. Some growers opt for techniques like topping or low-stress training to shape their plants for maximum yield. Others may adjust their lighting schedules to initiate flowering when the time is right

In closing, this book has taken you on a comprehensive journey through the art and science of cannabis cloning. From selecting the perfect mother plant to nurturing your clones through each crucial stage of growth, you've gained the knowledge and skills to cultivate healthy, thriving cannabis plants. Remember, successful cloning is not just a technique; it's an art that combines patience, care, and an understanding of the unique needs of your plants. With this newfound expertise, you are well-equipped to embark on your own journey of cannabis cultivation, creating a flourishing garden of your favorite strains.

Appendices

Appendix A: Common Pests and Diseases

Common pests and diseases that can affect cannabis plants during the cloning process and throughout their growth include:
Common Pests:
- Aphids: Small, soft-bodied insects that feed on plant sap, causing leaves to curl and become distorted.
- Spider Mites: Tiny arachnids that create fine webbing on plants and suck sap, leading to yellowing and stippling of leaves.
- Whiteflies: Small, white insects that feed on plant sap and can transmit plant viruses. They leave sticky honeydew on leaves.
- Thrips: Tiny, slender insects that scrape plant cells and feed on the sap, causing silvering or stippling of leaves.
- Fungus Gnats: Small, flying insects that lay eggs in the soil.

Larvae feed on roots, leading to poor growth.
- Caterpillars: Larvae of various moth species that can chew on leaves and damage plants.
- Scale Insects: Tiny, often immobile insects that attach themselves to plant stems and leaves, sucking plant juices.
- Mealybugs: Soft-bodied insects covered in a waxy substance that feed on plant sap.

Common Diseases:
- Powdery Mildew: A fungal disease that appears as a white, powdery substance on leaves and stems. It can inhibit photosynthesis and reduce yield.
- Botrytis (Gray Mold): A fungal disease that causes gray, fuzzy mold on buds and flowers. It can lead to bud rot and reduced quality.
- Root Rot: A condition caused by various fungi that affects the roots, leading to poor nutrient absorption and plant wilting.

- Bacterial Wilt: A bacterial disease that can cause wilting, yellowing, and a general decline in plant health.
- Leaf Septoria: A fungal disease that results in yellow or brown spots on leaves, often with a defined border.
- Leaf Spot: Various fungi can cause circular or irregular-shaped lesions on leaves, which can lead to reduced photosynthesis.
- Pythium (Damping-Off): A water mold that can cause seedlings and cuttings to rot at or near the soil line.
- Viral Diseases: Various viruses can affect cannabis plants, causing symptoms such as mosaic patterns on leaves, stunted growth, and leaf curling.

It's essential for cannabis growers to be vigilant in monitoring their plants for signs of these pests and diseases, as early detection and appropriate treatment are crucial for plant health and successful cloning. Integrated pest management (IPM) strategies, such as using beneficial insects, maintaining proper environmental conditions, and practicing good hygiene, can help prevent and manage these issues.

Appendix B: Cloning Progress Calendar

Month/Year: [Insert Month/Year]
Clone ID | Cutting Date | Transplant Date | Observations
- [Clone ID] | [Cutting Date] | [Transplant Date] | [Observations]Appendix C: Cloning Record-Keeping Templates

Instructions:
- In the "Month/Year" field, enter the current month and year for reference.
- For each row, fill in the following information:
  - Clone ID: Assign a unique identifier or name to each clone.

- Cutting Date: The date when the cutting was taken from the mother plant.
- Transplant Date: The date when the clone was transplanted into a larger container or growing medium.
- Observations: Record any relevant observations, such as growth rate, overall health, or specific issues encountered.
- As you progress through the cloning process, update the calendar to reflect the latest information for each clone.
- Use this calendar to track the development of your clones, identify any issues that may arise, and ensure that you provide appropriate care at each stage.

Having a visual record of your cloning progress can help you make informed decisions and optimize the success of your cannabis clones.

This section includes a collection of templates for record-keeping related to cannabis cloning. Templates include:

- Clone Identification Sheet: Record essential information about each clone, including strain, cutting date, mother plant source, and more.
- Environmental Log: Monitor and log environmental conditions, such as temperature, humidity, and lighting, to ensure optimal growing conditions.
- Nutrient and Feeding Schedule: Keep track of the nutrients and supplements used during the cloning process, along with feeding schedules.
- Pest and Disease Log: Document any pest or disease issues, including the date of discovery, symptoms observed, and treatment administered.
- Harvest and Yield Tracker: For those transitioning from clones to mature plants, use this template to record harvest dates and yield data.

Glossary

Glossary of Cannabis Cloning Terms

1. Cloning: The process of reproducing genetically identical copies of a cannabis plant, known as clones, from a mother plant.

2. Cutting: A vegetative branch or shoot from the mother plant that is taken for cloning. It includes a section of stem with one or more nodes.

3. Hardening Off: Gradually acclimating clones to outdoor conditions by exposing them to increasing amounts of natural light and airflow before transplanting them outdoors.

4. Harvest: The final stage of cannabis cultivation, when mature buds are ready to be harvested, dried, and cured for consumption.

5. Hygrometer: A tool used to measure humidity levels in the growing environment.

6. Low-Stress Training (LST): A method of gently bending and tying down branches to manipulate the shape of the plant and promote even light distribution.

7. Mold: Fungal growth that can develop on cannabis plants, often as a result of high humidity or poor ventilation.

8. Mother Plant: A mature, healthy cannabis plant selected for cloning due to its desirable traits, such as high potency, yield, or specific flavors.

9. Node: The point on a cannabis stem where a leaf connects. Nodes are crucial for root development and are often the location where roots emerge on cuttings.

10. Photoperiod: The duration of light and dark periods a cannabis plant receives in a 24-hour cycle. Photoperiod plays a role in controlling when plants enter the flowering stage.

11. Photoperiod Switch: Adjusting the light cycle to encourage cannabis plants to transition from the vegetative phase to the flowering phase.

12. Pruning: The act of selectively removing specific parts of a cannabis plant, such as leaves or branches, to manage growth and optimize plant health.

13. Pests: Insects or small organisms that can infest and damage cannabis plants, such as aphids, spider mites, or whiteflies.

14. Rooting Hormone: A substance, either in gel, powder, or liquid form, applied to the cut end of a clone to stimulate root growth.

15. Topping: A pruning technique that involves cutting off the top growth of a cannabis plant to encourage bushier growth and increase yields.

16. Transpiration: The process through which plants lose moisture vapor through small openings in their leaves (stomata).

17. Transplanting: The process of moving clones from their initial cloning environment to larger containers or growing mediums to support further growth.

18. Vegetative Phase: The growth stage in which cannabis plants focus on developing roots and green foliage, as opposed to flowering and bud production.

Frequently Asked Questions

1. FAQ: What is cannabis cloning, and why is it important?

- Answer: Cannabis cloning is the process of creating genetically identical copies (clones) of a mother cannabis plant. It's crucial for preserving desirable traits, ensuring consistency, and avoiding the uncertainty of growing from seeds.

2. FAQ: How do I select the right mother plant for cloning?

- Answer: Choose a mature plant with desirable traits like high potency, good yield, and disease resistance. Ensure it's pest-free and in excellent health.

3. FAQ: What's the best time to take cuttings from a mother plant?

- Answer: Take cuttings during the vegetative growth stage, typically before the plant enters the flowering stage. This stage promotes rapid root development.

4. FAQ: What are nodes, and why are they essential in cloning?

- Answer: Nodes are points on the stem where leaves or branches grow. They contain crucial meristematic tissue necessary for root development when cloning.

5. FAQ: What is rooting hormone, and do I need to use it?

- Answer: Rooting hormone is a substance that promotes root growth. It's recommended but not mandatory. It can significantly increase cloning success rates.

6. FAQ: How long does it take for clones to root?

- Answer: Clones typically root in 1-3 weeks, depending on factors like environmental conditions and the health of the mother plant.

7. FAQ: Can I use tap water for misting and watering clones?

- Answer: Using tap water is generally fine if it's low in chlorine and contaminants. However, some growers prefer distilled or filtered water to minimize risks.

8. FAQ: When is the best time to transplant clones into larger containers or the garden?

- Answer: Transplant clones when they have established a healthy root system, typically after 1-3 weeks. Ensure they have sufficient root mass before moving them.

9. FAQ: What's the ideal humidity level for cannabis clones?

- Answer: Maintain humidity around 70-75% for clones. High humidity helps prevent moisture loss and supports root development.

10. FAQ: Can I use artificial lighting for my clones, and if so, what type is best?

- Answer: Yes, you can use artificial lighting. Fluorescent or LED grow lights are suitable for clones, providing the right spectrum and intensity without causing stress.

11. FAQ: How often should I check my clones for pests and diseases?

- **Answer:** Regularly inspect clones, ideally daily, for signs of pests, diseases, or stress. Early detection and intervention are crucial.

12. FAQ: Do I need to adjust the light cycle for my clones, or can I keep them on the same schedule as my mother plant?

- **Answer:** Clones can remain on the same light cycle as the mother plant during the vegetative stage. Adjust lighting when transitioning to flowering.

13. FAQ: Should I prune or train my clones during the vegetative stage?

- **Answer:** Some light pruning or low-stress training (LST) can help shape clones for optimal growth and light distribution without causing stress.

14. FAQ: How long should I keep clones in the vegetative stage before switching to flowering?

- Answer: Keep clones in the vegetative stage until they reach the desired size, typically 4-8 weeks, depending on your goals and available space.

15. FAQ: Can I clone autoflowering cannabis strains?

- Answer: While it's possible to clone autoflowering strains, it's less common and may not result in significant time savings, as autoflowers have a predetermined life cycle. It's often more practical to grow them from seeds.

www.ingramcontent.com/pod-product-compliance
Lightning Source LLC
Chambersburg PA
CBHW062300290526
45794CB00006B/2633